AR 8.0/1.0

AR 8.0/1.0

20TH CENTURY SCIENCE AND TECHNOLOGY

1990-2000

THE ELECTRONIC AGE

Please visit our web site at: www.garethstevens.com
For a free color catalog describing Gareth Stevens' list of high-quality books and
multimedia programs, call 1-800-542-2595 (USA) or 1-800-461-9120 (Canada).
Gareth Stevens Publishing's Fax: (414) 332-3567.

Library of Congress Cataloging-in-Publication Data

Parker, Steve.
    1990-2000: the electronic age / by Steve Parker. — North American ed.
       p. cm. — (20th century science and technology)
    Includes bibliographical references and index.
    ISBN 0-8368-2947-6 (lib. bdg.)
    1. Technological innovations—History—20th century—Juvenile literature.
2. Inventions—History—20th century—Juvenile literature. [1. Technological
innovations. 2. Inventions—History—20th century. 3. Science—History—
20th century. 4. Technology—History—20th century.] I. Title.
    T173.8.P363   2001
    509.049—dc21   2001020786

This North American edition first published in 2001 by
Gareth Stevens Publishing
A World Almanac Education Group Company
330 West Olive Street, Suite 100
Milwaukee, WI 53212 USA

Original edition © 2000 by David West Children's Books. First published in Great Britain
in 2000 by Heinemann Library, Halley Court, Jordan Hill, Oxford OX2 8EJ, a division
of Reed Educational and Professional Publishing Limited. This U.S. edition © 2001 by
Gareth Stevens, Inc. Additional end matter © 2001 by Gareth Stevens, Inc.

Designers: Jenny Skelly and Aarti Parmar
Editor: James Pickering
Picture Research: Brooks Krikler Research

Gareth Stevens Editor: Dorothy L. Gibbs

Photo Credits:
Abbreviations: (t) top, (m) middle, (b) bottom, (l) left, (r) right

BMW: page 15(t).
Castrol: page 13(b).
Club Med: pages 14-15.
Corbis: cover (br), pages 4(m), 4-5(b), 12-13, 13(m), 14(b), 17(b), 18-19(b), 19(b), 21(both),
    24(b), 25(br), 26(l), 26-27.
Corbis Images: pages 4-5(t), 5(m), 16(both), 16-17, 28(t), 29(m).
Freeplay: pages 5(t), 20(b).
Glaxo: page 6(b).
NASA: pages 8(both), 8-9(both), 9, 10(both).
Rex Features: pages 6(t), 12(both), 17(m), 19(t), 22-23, 28(br).
Science Photo Library: page 11(b).
Sharp: page 22(b).
Sinclair: page 15(b).
Frank Spooner Pictures: cover (m), pages 5(b), 13(t), 17(t), 18-19(t), 23(b), 24(t), 25(bl), 27(b), 29(b).
David West: pages 7(t), 14(t), 18(b), 20(t), 23(m), 26(b).

Printed in the United States of America

1 2 3 4 5 6 7 8 9 05 04 03 02 01

20TH CENTURY SCIENCE AND TECHNOLOGY

1990-2000

THE ELECTRONIC AGE

Steve Parker

Gareth Stevens Publishing
A WORLD ALMANAC EDUCATION GROUP COMPANY

CONTENTS

INTO THE 21ST CENTURY 5

A NEW REALITY 6

SPACE SECRETS 8

WORLD SCIENCE 10

ON THE MOVE 12

GREEN TRAVEL 14

GENE GEOGRAPHY 16

SMART PARTS 18

TECH-TRONICS 20

SCREENLAND 22

DIGITS AND DISCS 24

GAMES GALORE 26

ULTRAMODERN MEDICINE 28

TIME LINE 30

GLOSSARY 31

BOOKS AND WEB SITES 31

INDEX 32

Space scientists are fairly certain that, except for Earth, life exists on no other planet in our solar system. They are not so sure, however, about life on moons, such as Saturn's Titan.

With many new engineering methods and materials, skyscrapers are getting taller and taller.

INTO THE 21ST CENTURY

Science and technology have affected daily life more during the last one hundred years than in the previous five thousand years. In developed nations, most people have cars, computers, mobile phones, and household appliances, such as microwave ovens and dishwashers. We watch live, global television and have instant Internet access to the full range of human knowledge. Yet science is a double-edged sword. It has produced nuclear weapons, global warming, acid rain, smog, and other dangerous pollution problems. As piles of radio-active waste grow, vast areas of natural beauty are destroyed. Despite medical successes and agricultural progress, one quarter of the world's people are still diseased or hungry. Will we use science and technology more wisely and fairly in the 21st century?

The windup radio combines high-tech electronic circuits with a low-tech clockwork motor.

Discovering the secrets of human genes will hopefully lead to many new medical treatments and cures.

"Eco-cars" use fewer natural resources to make, need less energy to run, reduce pollution, and take up less road space.

"Electronic pets" were a toy craze in the mid 1990s. They had to be fed and cared for, or they would become ill and die.

5

6

A NEW REALITY

The 20th century began with Einstein's mind-bending ideas on space, time, and relativity. It ended with even more amazing notions of infinite universes, time travel, cyberspace, and the nature of reality itself.

ALL IN THE MIND

Virtual reality (VR) makes an object or scene seem real to our senses — but it is not real. The object or scene does not exist as a physical entity. It exists only in the mind as make-believe. The brain builds an impression of reality from information fed to the body's senses by computer-controlled devices, such as a headset with earphones to make sounds and small screens to show pictures. These devices trick the senses and the brain into perceiving the unreal as real.

Using VR, an auto engineer can design and build a new engine, but it is only an image in a headset, not a physical object.

LIVING IN A VIRTUAL WORLD

VR took off in the 1990s as computers advanced to "real-time" processing, which means analyzing huge amounts of information and generating changing sounds, images, and movements at the same speed we experience them in actual life. VR is used in countless ways; for example, to train pilots, surgeons, and firefighters and to design cars, clothing, and shopping malls.

A medical scientist can manipulate a virtual model of a new drug molecule to test it and find out if it is effective.

Internet computers are linked by wires, optical fibers, radio signals, microwaves, and satellites. Gateways are major computer centers for routing information.

satellite

subnet

gateway

subnet

gateway

Apple iMacs were designed to send and receive data along wires rather than on disks.

"NET" POWER

The Internet, a global system of interconnected computers, has grown at an astonishing rate. In the late 1990s, the number of users doubled every six months. As more computers and information are added, the time may come when the "Net" has access to every scrap of knowledge we have — and might even become able to develop intelligence!

7

a closer part of the Universe

The negative energy associated with a wormhole creates a tunnel between space-time regions.

a very distant part of the Universe

CHEATING SPACE AND TIME

A wormhole is a region of space and time that has negative energy and negative mass. It can serve as a tunnel between two distant parts of the Universe, allowing positive energy, such as light, or even objects to pass through it, traveling billions of miles (kilometers) in an instant.

Two space-time regions distort.

The regions grow toward each other.

The throat of a wormhole opens.

The throat closes after a split second.

The regions separate again.

SPACE SECRETS

The 1990s was a perplexing time for space science. The "Red Planet," in particular, made headlines. Microscopic "sausages" found in a chunk of Martian rock were even hailed as signs of alien life.

A NEW VIEW

In 1990, the United States launched the Hubble Space Telescope (HST) into orbit around Earth. Far above the hazy, dusty atmosphere, the telescope should have had a clear view into space. Unfortunately, its mirror had not been polished to an accurate curve, so the view was blurred. In 1993, Space Shuttle astronauts fitted the HST with "contact lenses" to correct the problem.

Space Shuttle astronauts added more new equipment to the Hubble Space Telescope in 1997. The telescope was held near the open doors of the Shuttle's main cargo bay. The Shuttle's wing tips are showing on both sides of it.

The Hubble Telescope is 43 feet (13 meters) long, about the size of a 52-seat bus. It orbits 381 miles (613 km) above Earth's surface.

Russian space station Mir, *launched in 1986, suffered several accidents in the 1990s. In June 1997, an unmanned craft, ferrying supplies and taking away wastes, bumped into it. By 1999,* Mir *was mothballed in orbit, but a new, bigger space station was being assembled.*

ALIEN SAUSAGES?

In 1993, a fist-sized piece of rock collected by geologists in Antarctica, in 1984, was studied and found to come from Mars. It was probably blasted from the surface when a large meteorite crashed into the planet. The rock itself became a tiny meteorite for 15 million years and fell to Earth 13,000 years ago. Microscopes showed tiny, tubelike shapes in the rock. Were they the long-preserved remains of Martian microbes?

a piece of Mars with its tubelike microfossils

The HST has taken many awe-inspiring images of deep space that are more than a hundred times clearer than images from telescopes on the ground. This image is a small part of the Lagoon Nebula.

MYSTERIOUS MARS

The U. S. space probe *Mars Observer* visited Mars, Earth's nearest planet neighbor, in 1993. At least, it was supposed to. Radio contact with the probe was lost on its approach — a costly failure at $9 billion. Success, however, came in 1997. The probe *Mars Pathfinder* landed on Mars and sent back amazing live TV images as its wheeled, skateboard-sized rover vehicle, *Sojourner,* rolled across the planet's soft red dust. *Sojourner* moved at about $1/2$ inch (1 centimeter) per second, guided by its own lasers and other sensors and by an operator on Earth. A manned mission to Mars has been delayed, probably until 2020.

WORLD SCIENCE

In 1984, U. S. president Ronald Reagan announced a major new space project — a giant space station, code-named "Freedom." By the mid 1990s, after numerous cutbacks and delays, this project had become the International Space Station (ISS).

The main modules for the ISS form its laboratories. The larger units are 28 feet (8.5 m) long.

A JOINT VENTURE

The cost of the ISS is so great that many nations, including the United States, Japan, Russia, Canada, and some European countries, are involved in its development. The ISS is being assembled in orbit, piece by piece, by means of forty-four rocket launches. In the year 2000, the original completion date of 1995 was moved out to 2005.

An artist's depiction of the ISS shows solar panels the size of a soccer field. The station would weigh 400 tons on Earth's surface, but in orbit, it is weightless.

THE VALUE OF THE ISS

Why spend billions on the ISS when so many people could benefit from that money? Practical reasons include the lack of gravity, which means that super-pure substances, such as crystals for microchips, can be made in orbit. There is also the prospect that new discoveries in deep space might change our ideas about the origin and fate of the Universe.

A DISTANT WORLD

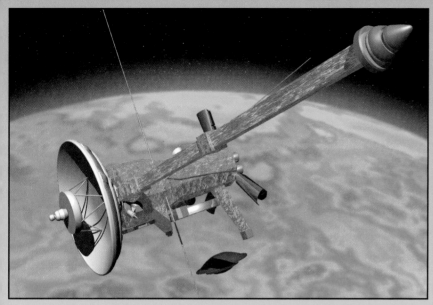

In its Saturn orbit, the Cassini-Huygens probe will separate.

Named after Giovanni Cassini (1625–1712), who discovered four of Saturn's moons and the gap in its rings, the Cassini orbiter will study that planet's atmosphere. The Huygens lander, which was named after Christiaan Huygens (1629–1695), who discovered Saturn's rings and its moon Titan, will detach from Cassini and parachute toward Titan to land. Huygens will radio information about Titan's surface and gases to Cassini, which will relay the reports to Earth.

Its atmosphere-entry dish will protect the Huygens lander.

Its parachute will open at a height of 106 miles (170 km).

The lander will settle on Titan.

PROBING SATURN

The Cassini-Huygens space probe was the last major unmanned mission of the 1990s. It set off in 1997 to reach the giant outer planet Saturn in 2004. The Cassini orbiter will circle Saturn, studying its beautiful rings. The Huygens lander will approach the planet's largest moon, Titan. Some scientists believe Titan may once have had conditions suitable to support life — at least, life as we know it.

THE MICROROCKET

Big rocket engines for space travel are heavy, expensive, and thirsty for fuel. The microrocket is only 0.12 inches (3 millimeters) thick and 0.6 inches (15 mm) wide, yet it is twenty times more efficient than a big rocket engine. Burning methane and oxygen gases for fuel, approximately 800 microrockets could launch a one-ton craft into space. Like an electronic microchip, the microrocket is made from a wafer of silicon.

A microrocket is smaller than the tip of a thumb.

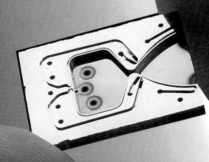

ON THE MOVE

After the boom of the 1960s and the bust of the 1970s and 1980s, travel and transportation settled down in the 1990s. There was still room, however, to be big, fast, and expensive.

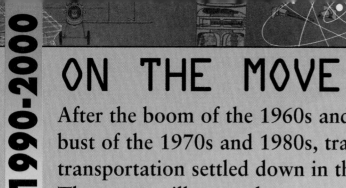

With a Global Positioning System (GPS), travelers can know their locations to within a few yards (m).

A new type of maglev train tested in Japan in 1997 reached speeds of about 500 miles (800 km) per hour.

Japan's ultrastreamlined, electric intercity trains travel as fast as some passenger aircraft.

MASS TRANSIT

In Japan, a third generation of sleek "bullet trains" reached speeds of 186.5 miles (300 km) per hour. These trains run on the usual system of wheels and rails, but the tracks must be very straight and supersmooth. Maglev (magnetic levitation) trains, suggested as far back as the 1930s, were still being explored in the 1990s, but they made little headway. These trains float above a track, held up by magnetic force.

Radio signals from satellites and local ground beacons can identify a car's exact location. They can even help a driver operate the car so it follows the correct route to a preprogrammed destination.

Antenna transmits and receives satellite and beacon signals.

engine control unit

steering and accelerator feedback

dashboard display

onboard computer

RECORD BREAKERS

Except for space rockets, the most expensive aircraft ever built was the Northrop B-2 Spirit "stealth bomber." This craft first flew in 1989 and was delivered to the U. S. Air Force in 1993 — with a price tag of some $600 million. Its batlike design has curves, zigzag edges, body coverings, and special paint to make it undetectable by enemy radar. The B-2 was proposed in the late 1970s, during fear of a nuclear war between the United States and the Soviet Union. U. S. plans to build 132 B-2s changed in 1990 to about twenty.

13

The B-2 is powered by four turbofan jet engines. It has a wingspan of 172 feet (52.4 m) and a top speed of 470 miles (760 km) per hour.

In 1997, the Thrust SSC (supersonic car) set a land speed record of 762.5 miles (1,227 km) per hour and became the first car to break the sound barrier.

GREEN TRAVEL

In the 1990s, vehicle designers took more notice of the public's concerns about the environment. New models were "greener" in that they wasted less fuel and precious natural resources, produced less pollution, and could be recycled more easily.

Modern Mini cars employ existing technologies made smaller. They use less fuel and raw materials and take up less road space.

Club Med 1 is a luxury cruise ship that can use sail power to save engine fuel.

CAR PROBLEMS

Cars offer the freedom to travel when and where we wish, with our passengers and luggage protected from the weather. A car, however, takes huge amounts of materials and energy to manufacture. Electric cars are kinder to the environment, with motors that are 90 percent efficient compared to a gasoline or diesel motor at 30 or 40 percent efficient.

Electric cars rely on batteries that are conveniently recharged overnight.

14

RECYCLING A CAR

An old car on a scrap heap is a huge waste of natural resources. New guidelines in various regions specify how much of a car should be made from recycled materials, especially plastics, and which parts should be recyclable so they can be used again in the future.

▮ *plastic parts made from recycled materials*

▮ *plastic parts made to be recycled in the future*

BMW 3 series

COMBINED TECHNOLOGIES

Innovative projects are combining past and present technologies. For ships, low-tech wind power is free, but high-tech engine power is reliable and convenient. Some new ships have both. A shipboard computer analyzes the wind's speed and direction, then calculates whether fuel can be saved by raising the sails.

ALTERNATIVES

Technology has developed some "green" alternatives to cars, such as the electric bicycle and faster, more comfortable types of public transportation. People, unfortunately, are still very attached to the convenience of their own vehicles. It is more a problem for society than science to change people's ideas and expectations.

◉ FINE-WEATHER TRAVEL ◉

The greenest, healthiest, and most energy-efficient way to make short trips is by bicycle. In cold, wet places, however, cycling can be an awkward way to travel. One attempt to make cycling more attractive is the electric bicycle. Rechargeable batteries in its frame power an electric motor, but it also has normal pedals for extra push.

the Zike electric bicycle

GENE GEOGRAPHY

The 1990s was the decade of gene power. The genetic information of living things was read, engineered, modified, duplicated, and patented. With the new millennium came a tremendous achievement — knowing all of our own human genes.

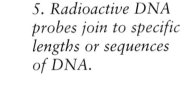

GENETIC FINGERPRINTS

Genes consist of very long, twisted chemicals called DNA, or deoxyribonucleic acid. Like any chemical, DNA can be isolated, made pure, and altered in a laboratory. A genetic fingerprint identifies an individual's DNA.

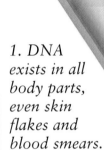

1. DNA exists in all body parts, even skin flakes and blood smears.

2. The DNA is cut into pieces.

3. A method known as electrophoresis separates DNA pieces into bands.

4. The DNA bands are put onto a nylon sheet.

5. Radioactive DNA probes join to specific lengths or sequences of DNA.

6. The end result is a unique DNA bar code, or fingerprint.

THE HUGO PROJECT

The Human Genome Organization (HUGO) was set up in 1990 to unravel the 100,000-plus genes in the human genome. Carried out mainly by governments and universities, the project encountered problems that pushed its completion date beyond 2005. When private genetic companies got involved, the "gene map" was completed by 2000.

This lab worker is separating DNA samples.

Small, dark bars indicate certain DNA sequences or codes that are parts of genes.

16

In 1997, Dolly the sheep made headlines as the first identical genetic copy, or clone, made from the cells of another adult mammal.

The tiniest sample of a living thing can yield DNA material to make a clone or a fingerprint.

ULTIMATE IDENTITY CARD

Genetic fingerprints have been used since the 1980s to identify people from body samples as tiny as a single hair or a drop of saliva. Knowing the human genome is another major step forward. It is the complete set of instructions for developing and maintaining a human body. The huge task remains, however, to find out exactly how each gene works and if genes can be "mended" to cure genetic diseases.

THE GM DEBATE

In the 1990s, genes of farm crops, including tomatoes, corn, and soy, were altered or modified. For example, a gene was added to protect a plant against certain weed killers so fields could be sprayed more effectively. Some people worried that modified genes might swap with natural genes of wild plants and animals and, once in the environment, could never be removed.

Genetically modified (GM) crops and farm animals could help global food shortages (above), but protesters feared genes might "escape," so they disrupted the GM crop tests (left).

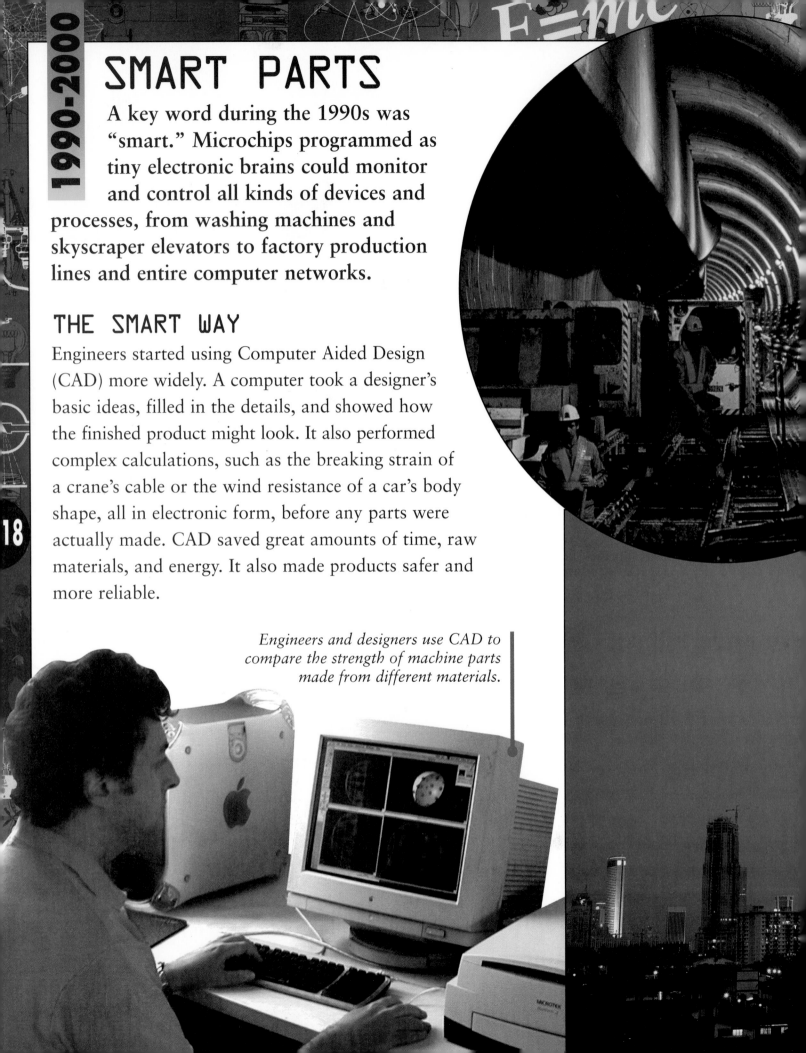

SMART PARTS

A key word during the 1990s was "smart." Microchips programmed as tiny electronic brains could monitor and control all kinds of devices and processes, from washing machines and skyscraper elevators to factory production lines and entire computer networks.

THE SMART WAY

Engineers started using Computer Aided Design (CAD) more widely. A computer took a designer's basic ideas, filled in the details, and showed how the finished product might look. It also performed complex calculations, such as the breaking strain of a crane's cable or the wind resistance of a car's body shape, all in electronic form, before any parts were actually made. CAD saved great amounts of time, raw materials, and energy. It also made products safer and more reliable.

Engineers and designers use CAD to compare the strength of machine parts made from different materials.

The 31-mile (50-km) Channel Tunnel between France and England is the world's longest undersea tunnel. It opened in 1994 after seven years of construction.

Using nanotechnology, complex machines, such as an electric motor with a gearbox, can be made smaller.

The 88-story Petronas Towers are 1,483 feet (452 m) tall. A skywalk links the twin towers at floor 42.

TIGER TECHNOLOGY

In the 1990s, an increasing rate of technical innovation in Southeast Asia led to rapidly expanding economies in those countries. Their so-called "tiger economies" and their adaptable workforces attracted money from rich Western nations to develop all kinds of products, from microchips to oil tankers. The Petronas Towers were symbols of this growth. Completed in 1997, these twin skyscrapers in Kuala Lumpur, Malaysia, replaced Chicago's Sears Tower as the world's tallest building.

The "tiger economies" of Southeast Asia provided millions of jobs in manufacturing industries, especially in electronics.

TECH-TRONICS

During the microchip revolution of the 1990s, electronic devices became faster and more powerful, yet smaller and less expensive. One of the most noticeable results was the mobile phone. In just a few years, this communication innovation transformed from a bulky rarity into a palm-sized, "must-have" accessory.

With a new-generation mobile phone and a laptop computer, you can access the Internet from almost anywhere.

OLD AND NEW TECHNOLOGIES

The "windup radio" was developed for use in areas where electricity and batteries are not readily available. It is driven by human muscle power. After the radio is wound up, its old-fashioned clockwork mechanism unwinds slowly. As it unwinds, it spins a small electricity generator that powers its modern electronic circuits. A windup radio can usually run 1 to 2 hours on a 1- or 2-minute winding.

The Freeplay windup radio and its inventor, Trevor Baylis, have won many awards.

MORE AND MORE MOBILES

In 1997, Finland became the first country in the world to have more mobile phone numbers than ordinary, fixed-line phone numbers. By 2000, more than half the people in Britain had mobile phones. During the 1990s, the size and weight of mobile phones were reduced by four-fifths, despite the addition of new features such as 100-number phone book memories, voice mail, text messages, and advanced WAPs (wireless application protocols) to allow more types of links and Internet access.

OFFICE ON THE MOVE

Portable computers in the 1990s were also made smaller and lighter, yet faster and more powerful. Several different types appeared. Each type was specialized for certain jobs. A typical palmtop runs for hours on a single battery charge and uses powerful programs, from word processing to graphics, but it is small enough to fit into a pocket. It has a docking port for connecting it to a larger computer, so work done on the move can easily be transferred. A notebook computer is specialized mainly for word processing.

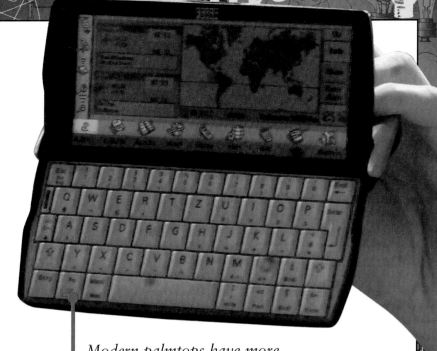

Modern palmtops have more functions than earlier models, including access to the Internet.

NEVER GET LOST AGAIN!

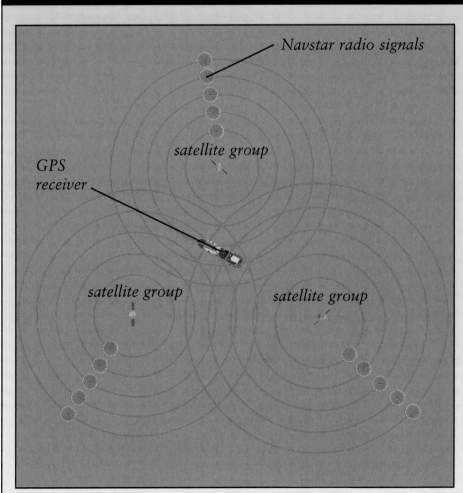

Navstar radio signals

satellite group

GPS receiver

satellite group

satellite group

In the Global Positioning System (GPS), which became fully operational in 1994, three groups of Navstar satellites are in different orbits around Earth. Each of a group's seven satellites sends out radio messages with its own identification code, its position, and a very accurate time signal. Anywhere on Earth, at least three satellites are within the range of a GPS receiver, which detects the satellites' signals and compares the times they took to arrive, determining the distance to each satellite. The GPS pinpoints a receiver's position to within a few yards (m).

A handheld GPS receiver displays a location with either map coordinates or an actual map of the area on its screen.

SCREENLAND

One worldwide survey estimates that, between 1990 and 2000, the number of screens in the world increased by at least a hundred times. More and more, people are looking at images on a screen instead of at the real world.

SCREENS ARE EVERYWHERE

Screens of many kinds are now found almost everywhere, not just on television sets and computer monitors. Screens are on video cameras and on data displays, such as those in control rooms. They are also on aircraft flight decks, on camera security systems, and on medical scanners in hospitals. Even at stadium events, giant screens give people at the back of the crowd a better view.

A computer is the central component of a growing system that can be used for work or play — games, listening to music, watching TV, or surfing the Internet.

A video camera screen lets the user see what has been recorded as if it were on a television screen. The image can then be either kept or erased.

SCREEN TECHNOLOGY

There are two basic types of screen technology. The boxlike CRT (cathode ray tube) screen was first developed in the 1930s. Its technology fires electron beams, formerly known as cathode rays, at the back of a glass screen, making tiny dots of phosphorescent chemicals on the glass glow. Flat screens are much thinner and lighter and use less electricity than CRTs, but they usually are not quite as bright, sharp, and clear. They use LCD (liquid crystal display) technology. Tiny, crystal-like units block light or let it through depending on the electrical signals they receive.

During the 1990s, the number of people working from home increased by about five times. They used a computer system linked by telephone lines to colleagues, e-mail systems, and the Internet. A growing variety of electronic devices can be added to a basic computer.

digital camera — *monitor* — *Internet connection*

palmtop

MP3 music player — *television* — *speaker* — *printer*

scanner

removable disk drive — *keyboard* — *CD drive* — *mouse*

Apple's iMacs did not come with standard removable disk drives. They were meant to communicate along wires.

THE VERSATILE CD

Compact discs (CDs) were introduced in the early 1980s, mainly to store recorded music. CDs, however, are not limited to storing music. They can hold any kind of information, or data, in the same digital format — codes of microscopic pits in the surface that can be read by a tiny laser beam. In the early 1990s, many computers were equipped with CD drives so they could read or obtain data, such as programs, words, pictures, and sounds, from CDs. They could not, however, write or record information onto CDs. By the mid 1990s, many computers came with CD writers, or "burners."

23

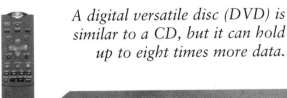

A digital versatile disc (DVD) is similar to a CD, but it can hold up to eight times more data.

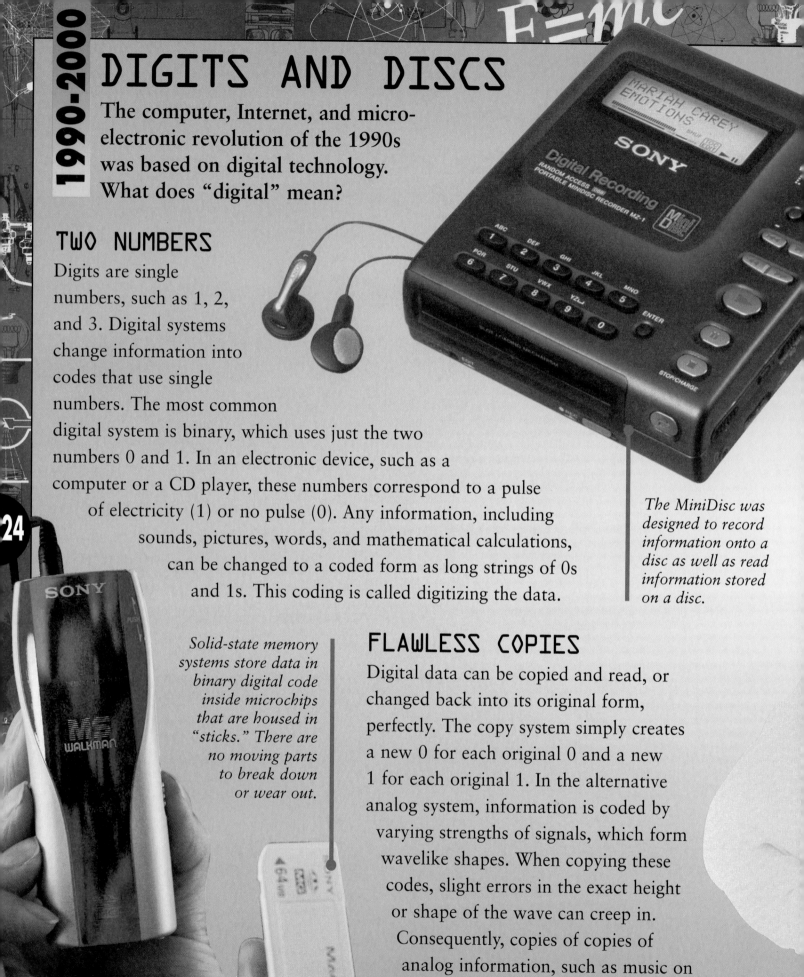

DIGITS AND DISCS

The computer, Internet, and micro-electronic revolution of the 1990s was based on digital technology. What does "digital" mean?

TWO NUMBERS

Digits are single numbers, such as 1, 2, and 3. Digital systems change information into codes that use single numbers. The most common digital system is binary, which uses just the two numbers 0 and 1. In an electronic device, such as a computer or a CD player, these numbers correspond to a pulse of electricity (1) or no pulse (0). Any information, including sounds, pictures, words, and mathematical calculations, can be changed to a coded form as long strings of 0s and 1s. This coding is called digitizing the data.

24

The MiniDisc was designed to record information onto a disc as well as read information stored on a disc.

Solid-state memory systems store data in binary digital code inside microchips that are housed in "sticks." There are no moving parts to break down or wear out.

FLAWLESS COPIES

Digital data can be copied and read, or changed back into its original form, perfectly. The copy system simply creates a new 0 for each original 0 and a new 1 for each original 1. In the alternative analog system, information is coded by varying strengths of signals, which form wavelike shapes. When copying these codes, slight errors in the exact height or shape of the wave can creep in. Consequently, copies of copies of analog information, such as music on magnetic tape, are gradually altered.

RECORD AND PLAY

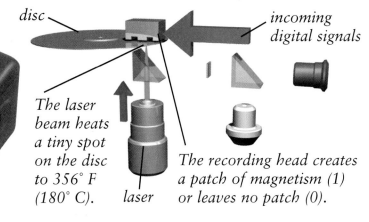

disc

incoming digital signals

path of the beam for no pulse (0)

The laser beam heats a tiny spot on the disc to 356° F (180° C).

laser

The recording head creates a patch of magnetism (1) or leaves no patch (0).

The laser beam reflects off the disc.

path of the beam for pulse (1)

beam sensor

outgoing digital signals

RECORDING (writing data)

PLAYING (reading data)

Sony's MiniDisc system, developed in 1992, uses two different technologies to make its compact discs re-recordable. First, a laser beam heats a tiny spot of the metal foil on the disc. Then a magnetic recording head swipes over it, leaving a tiny patch of magnetism. The magnetism is enough to alter a laser beam shining on the disc. The patch, detected as data, is a digital code 1. No patch is a 0.

PERSONAL TV

The first portable "pocket TVs" appeared in the 1960s, but detecting television signals while the user is moving around is a problem, and the picture or sound often fade. Modern tuning circuits can get around this problem in areas where TV signals are strong. The portable, bright-screen DVD-TV, which uses DVDs, is far more reliable.

This small DVD-TV plays movies and other programs stored on CD-sized DVDs, or digital versatile discs.

25

You can see through these "TV specs" even when a tiny projector shines images onto a screen that is on the inside of the eyepiece.

GAMES GALORE

The shrinking size and falling cost of microchips and other electronic devices created a whole new area of leisure-time fun — handheld electronic games, or e-games. A plug-in cartridge contains the software, or program, and a small screen on the handset displays the action.

BIGGER IS NOT ALWAYS BETTER

Early in the 1990s, arcade-style video games bigger than a person showed fast-moving, lifelike race cars, aliens, and other typical scenes. By the end of the decade, the same, or even better, speed and quality were available on game consoles people could hold in one hand. Some consoles plug into a TV set or a computer monitor. Others have built-in visual displays.

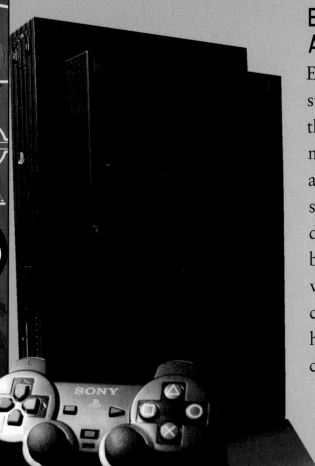

Nintendo's Game Boy, launched in 1989, remained hugely popular through the 1990s. Color screens and bright new cases kept it looking up-to-date.

LOOKING GOOD

How game consoles looked was almost as important as what they could do. Styles ranged from cool grays and metallic sheens in the early 1990s to bright, brash colors and see-through plastic casings a few years later.

Sony's PlayStation arrived in 1995. Its small box is said to contain almost as much computing power as used for the first Apollo Moon landing in 1969.

26

THE NEXT CHALLENGE

In the late 1990s, some game manufacturers tried to tackle mass-market virtual reality (MMVR). Producing bright, detailed images and clear sounds that change at the same speed as in real life requires great computing power. By the year 2000, microchips and other components were powerful and cheap enough for some MMVR systems to go on sale. There was a lot of debate, however, about their safety. Inside a VR headset, absorbed in computer-generated sights and sounds, the user is effectively cut off from the real world and may not be able to respond to a real-life emergency.

Home VR systems may be the next generation in e-games. Some experts think they will encourage certain users to retreat from real life into an isolated virtual world.

NEW IDEAS

The e-game craze of the 1990s was the virtual pet, a small electronic device that responded like a real pet. It had to be fed, watered, cleaned, exercised, and trained. It would get sick, or even die, if its owner did not take good care of it. Some people thought virtual pets were good training for having real ones.

Virtual pets, known as Tamagotchi, were developed in Japan. Besides being fun, they react like real pets for wishful pet owners who may not have the space or facilities for a live animal.

ULTRAMODERN MEDICINE

Medical science moved into the 21st century with more implantable parts, surgical equipment, lasers, scanners, and other high-tech devices than ever before. People have never been healthier or lived longer.

HIGH-TECH HUMAN

Prostheses are replacements for body parts. Some are "false," non-working versions, but electronic parts are also being developed.

1. composite skull plate
2. electronic light-sensitive eye
3. nose bridge
4. hearing aid implant
5. alloy jaw plate
6. plastic chin implant
7. electronic voice box
8. shoulder joint
9. artificial heart
10. elbow hinge
11. arm prosthesis
12. metal forearm plate
13. wrist bone replacement
14. thumb-wrist link
15. ball-and-socket hip joint
16. thigh bone implant
17. knee bridge
18. artificial leg
19. false big toe

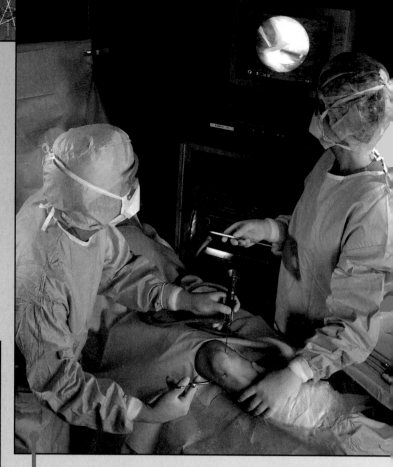

Many types of "scopes," made with flexible optical fibers, look into the human body. This arthroscope is being used to see the inside of a knee joint, an area often affected by sports injuries.

LONG-TERM BATTLES

Smoking has long been associated with heart and lung diseases, various forms of cancer, and many other illnesses. The human body can become dependent on the drug nicotine in tobacco smoke. Special patches provide an alternate source of nicotine that is supplied through the skin. They avoid the tremendous harm caused by tobacco smoke and can gradually be reduced in strength.

Nicotine patches supply the body with the drug it craves without damaging the lungs.

MENDING GENES

Faulty genes cause hundreds of diseases and medical conditions. Unraveling the full set of human genes has opened up the incredible possibility of gene therapy, which involves "mending" faulty genes at an early stage of life by replacing them with correct versions.

Tiny samples of body tissues can be examined in a laboratory to look for genetic problems that will cause trouble later in life.

Viruses are the smallest living things. Although many of them cause diseases, some can be helpful. They are used as carriers to move a correct gene into a human cell to replace a faulty one. The human cell then multiplies to cure the disease.

virus genetic material

Virus enters body cell where new gene joins human genetic material.

replacement gene

New gene is spliced into virus genes.

THE FUTURE

Twentieth-century advancements in science and technology, from heart transplants and high-yield farm crops to space travel and virtual reality, were amazing. Yet, in the 21st century, millions of people all over the world are still homeless, hungry, and suffering. The problem now is ensuring that achievements in science and technology are used to help everyone.

29

Vaccinations, or immunizations, to prevent the spread of serious infectious diseases are one of medicine's greatest triumphs. These Ethiopians are waiting to be immunized against polio.

- TIME LINE -

	WORLD EVENTS	SCIENCE EVENTS	TECHNOLOGY	FAMOUS SCIENTISTS	INVENTIONS
1990	• Gulf War breaks out as Iraq invades Kuwait	• Geneva: World Climate Conference	• Hubble Space Telescope launched • TGV Atlantique hits 320 miles (515 km) per hour	• Michael Irwin discovers a tenth small galaxy orbiting Earth's galaxy, the Milky Way	• "Toddler Tag" electronic radio bracelet for children • First International Robot Olympics at Strathclyde University, Glasglow
1991	• Breakup of USSR • Yeltzin leads Russia	• Suggestions that "hot, dark matter" giving out infrared radiation may hold the Universe together	• Dyson's bagless vacuum cleaner wins design prize • Pen-sensing electronic notepads developed	• Richard Ernst wins Nobel Prize for improvements to medical MR (magnetic resonance) scanners	• Pocket electronic encyclopedia
1992	• U.S.: race riots in Los Angeles • Australia drops oath of loyalty to Queen of England	• Rio de Janeiro: "Last Chance" Earth summit • COBE satellite detects echoes of the Big Bang	• Chris Boardman wins gold medal riding new-concept Lotus Superbike in Barcelona Olympics	• Galileo (died 1642) receives a formal apology from the Pope that Earth does indeed orbit the Sun	• Renault: Zoom folding electric car • Philips: Digital Compact Cassette (DCC) • Laser potato peeler
1993	• PLO and Israel sign peace agreement	• Shoemaker-Levy comet torn apart by Jupiter's gravity	• Movie special effects reach new heights with Stephen Spielberg's Jurassic Park	• Joseph Taylor and Russell Hulse receive Nobel Prize for discovery of binary pulsars	• Video phones • Apple: Newton • Casio: Alti-Thermo watch
1994	• South Africa: Mandela is first black president • Civil war in Rwanda	• Genes linked to breast cancer identified • Genetically modified tomatoes first sold	• Channel Tunnel completed • Number of Internet users exceeds 30 million	• Philip Gurner invents the laser-guided peashooter for World Peashooting Championships	• Zeta strap-on electrical power pack for bicycles
1995	• U.S.: terrorist bomb blast in Oklahoma City • Kobe earthquake • Tokyo: nerve gas attacks by Aum Shinrikyo cult	• Satellites show the Antarctica icecap is shrinking by 1.4 percent every 10 years	• "Liquid bone" (Norian SRS), which is injected and gradually changes to real bone, developed to stabilize fractures	• Harold Cohen of San Diego writes the computer program, Aaron, which develops its own art style	• Baylis: Freeplay windup radio
1996	• "Mad Cow" disease: bans on British beef	• Possible microbe fossils in meteorite from Mars • First antimatter, anti-hydrogen, made at CERN Centre in Europe	• UK: Sea Wraith (stealth ship prototype) • Number of Internet users reaches 50 million	• Smalley, Curl, and Kroto receive Nobel Prize for golfball-like "buckyball" molecules	• Sony: PlayStation • Benecol cholesterol-lowering margarine
1997	• UK returns Hong Kong to China • Roswell report denies alien encounter • UK: Blair named prime minister	• Adult mammal, Dolly the Sheep, cloned • World's oldest rocks (4,000 million years)	• Sojourner rover on Mars • Thrust SSC breaks sound barrier	• IBM's Deep Blue beats Kasparov at chess • Borge Ousland crosses Antarctica alone	• Bandai: Tamagotchi • Apple: eMate • Windup flashlight • LawnNibbler robotic lawn mower
1998	• South Africa: Truth and Reconciliation Report • Birth of the euro currency	• Construction of International Space Station begins	• VW "Beetle" gets a high-tech makeover for the new millennium	• U.S. heart surgeon Denton Cooley receives National Medal of Technology	• Apple: iMac • Digital versatile disc (DVD) • Smart micro car goes on sale in Europe
1999	• NATO air strikes on Yugoslavia • India and Pakistan: nuclear testing crisis • Serbians drive Albanians from Kosovo	• New artificial chemical element, number 114, created	• Japan builds largest single telescope mirror, 27 feet (8.2 m) across, in Hawaii	• Ahmed Zewail receives Nobel Prize for taking billion-billionth-second photos of molecules	• Zanussi: Zoe washing machine, Oz refrigerator, and Teo oven • DreamCast computer games

GLOSSARY

analog: related to a mechanism, such as a computer system, in which data is represented by measurable and variable physical quantities, such as the amount of electrical resistance, and which usually presents solutions in a graphic form or shape, such as a wave.

compact disc (CD): a memory device consisting of a plastic disc with microscopic pits that digitally code information such as sounds, pictures, words, and computer programs.

electronic circuit: an assembly of electronic elements that form a complete path of electric current.

gene: one of many tiny elements, made of chemicals called DNA, that are part of a chromosome and which transfer inherited characteristics from parent to offspring.

human genome: the complete genetic material of a human organism.

laser: a device that produces a powerful beam of intense, high-energy, pure-color light. The term stands for Light Amplification by Stimulated Emission of Radiation.

nanotechnology: a technical system or process that manipulates the atoms and molecules of materials to build microscopically small devices.

nebula: a huge light or dark cloud in interstellar space, especially in the Milky Way galaxy, which is made of gases and contains small amounts of dust.

prostheses: artificial devices that replace missing parts of the human body.

virtual reality: an artificial environment in which computer-generated stimuli, such as sights and sounds, are perceived as real by the senses and the brain, and what happens in this environment is determined partly by the user's actions.

MORE BOOKS TO READ

The 90s: The Digital Age. 20th Century Design (series). Hannah Ford (Gareth Stevens)

Adventure in Space: The Flight to Fix the Hubble. Elaine Scott (Disney Press)

The Adventures of Sojourner: The mission to mars that thrilled the world. Susi Trautmann Wunsch (Firefly Books)

Biotechnology. Inventors & Inventions (series). Donna Koren Wells (Benchmark)

Cloning: Frontiers of genetic engineering. Megatech (series). David Jefferis (Crabtree)

E-Mail and Internet. In Touch: Communicating Today (series). Chris Oxlade (Heinemann Library)

Eyewitness: Electronics. Roger Bridgman (DK Publishing)

High Speed Trains. Built for Speed (series). Holly Cefrey (Children's Press)

Multimedia Magic. Computer Science (series). Robert L. Perry (Franklin Watts)

Virtual Reality: Experiencing Illusion. New Century Technology (series). Christopher W. Baker (Millbrook Press)

WEB SITES

Human Genome Project. *www.ornl.gov/ TechResources/Human_Genome/home.html*

Inside the Space Station: life in space. *www.discovery.com/stories/science/iss/iss.html*

MiniDisc: Facts and FAQs. *www.minidisct.com/md_facts.html*

Virtual Pet Home Page. www.virtualpet.com/vp

Due to the dynamic nature of the Internet, some web sites stay current longer than others. To find additional web sites, use a reliable search engine with one or more of the following keywords: *binary code, CAD, Cassini-Huygens, DVD (digital versatile disc), electric cars, genes, GPS (global positioning system), mobile phones,* and *virtual reality.*

INDEX

acid rain 5
agriculture 5
  crops 17, 29
aircraft 12, 13, 22
  Northrop B-2 Spirit
    stealth bomber 13
analog technology 24
animals 17, 27
Apple iMacs 7, 23

Baylis, Trevor 20
bicycle, electric 15

cameras 22
  digital 23
  video 22
cars 5, 6, 13, 14, 15,
  18, 26
  BMW 3 series 15
  eco-cars 5
  electric 14
  Minis 14
  Thrust SSC 13
Cassini, Giovanni 11
Cassini-Huygens
  probe 11
clones 17
  Dolly the sheep 17
compact discs (CDs) 23,
  24, 25
Computer Aided Design
  (CAD) 18
computers 5, 6, 7, 13,
  15, 18, 21, 22, 23,
  24, 26, 27
  games 22, 26, 27
    Game Boy 26
    PlayStation 26
  laptops 20
  notebooks 21
  palmtops 21, 23
cyberspace 6

digital technology 23,
  24, 25
  binary system 24
  codes 23, 24

digital versatile discs
  (DVDs) 23, 25
DNA 16, 17
drugs 6, 28

Einstein, Albert 6
electronic circuits 5,
  20, 25
electronic games 26, 27
electronic pets
  (*see* virtual pets)
electronics 11, 18, 19,
  20, 23, 24, 26, 27, 28
energy 5, 7, 14, 15, 18
engineering 4, 16
engineers 6, 18
engines 6, 11, 13, 14, 15
environment 14, 17

fuel 11, 14, 15

gene modification (GM)
  16, 17
gene therapy 29
genes 5, 16–17, 29
genetic fingerprints
  16, 17
Global Positioning
  System (GPS) 12, 21
global warming 5

Hubble Space Telescope
  (HST) 8, 9
Human Genome
  Organization
    (HUGO) 16
Huygens, Christiaan 11

International Space
  Station (ISS) 10
Internet 5, 7, 20, 21, 22,
  23, 24

lasers 9, 23, 25, 28

machines 18, 19
magnetic force 12

magnetism 25
manufacturing 14,
  19, 27
Mars Observer probe 9
Mars Pathfinder probe 9
medicine 5, 6, 28–29
meteorites 9
microchips 10, 11, 18,
  19, 20, 24, 26, 27
microfossils 9
microwaves 5, 7
Mir space station 9
mobile phones 5, 20
Moon landing, Apollo 26
moons 4, 11, 26
  Titan 4, 11
motors 5, 14, 15, 19
  diesel 14
  gasoline 14
music 22, 23, 24
  MP3 player 23

nanotechnology 19
natural resources 5,
  14, 15
nicotine patches 28
Nintendo Game Boy 26

optical fibers 7, 28

planets 4, 8, 9, 11
  Earth 4, 8, 9, 10,
    11, 21
  Mars 8, 9
  Saturn 4, 11
plastics 15, 26, 28
pollution 5, 14
prostheses 28

radio signals 7, 13, 21
radioactive waste 5
radios 9, 11, 20, 21
  windup 5, 20
Reagan, Ronald 10
recycling 14, 15
rockets 10, 11, 13
  microrockets 11

satellites 7, 13, 21
  Navstar 21
science 5, 8, 10, 15,
  28, 29
scientists 4, 6, 11
screens 6, 21, 22, 25, 26
  cathode ray tubes
    (CRTs) 22
  liquid crystal displays
    (LCDs) 22
ships 14, 15
  Club Med 1 14
skyscrapers 4, 18, 19
  Petronas Towers 19
  Sears Tower 19
Sojourner rover 9
Sony MiniDisc 24, 25
Sony PlayStation 26
space 4, 6, 7, 8–9, 10,
  11, 13, 29
Space Shuttle 8

Tamagotchi 27
technology 5, 14, 15,
  19, 22, 25, 29
  analog 24
  digital 23, 24, 25
television (TV) 5, 9, 22,
  23, 25, 26
time 6, 7, 18, 21
tobacco smoke 28
trains 12
  bullet 12
  electric 12
  maglev 12
transportation 12, 15
travel 6, 11, 12, 14,
  15, 29

vaccinations 29
vehicles 9, 14, 15
virtual pets 5, 27
virtual reality 6, 27, 29

wormholes 7

Zike electric bicycles 15

32